U.S. Department
of Transportation

Research and Innovative Technology
Administration

Volpe National Transportation
Systems Center

Final Report and Recommendations for Research on Human-Automation Interaction in the Next Generation Air Transportation System

NASA Airspace Systems Program

Final Report
November 2006

Thomas B. Sheridan, Kevin M. Corker and Eric D. Nadler

Notice

This document is disseminated under the sponsorship of the Department of Transportation in the interest of information exchange. The United States Government assumes no liability for its contents or use thereof.

Notice

The United States Government does not endorse products or manufacturers. Trade or manufacturers' names appear herein solely because they are considered essential to the objective of this report.

Final Report and Recommendations for Research on Human-Automation Interaction in the Next Generation Air Transportation System

REPORT DOCUMENTATION PAGE	Form Approved OMB No. 0704-0188

Public reporting burden for this collection of information is estimated to average 1 hour per response, including the time for reviewing instructions, searching existing data sources, gathering and maintaining the data needed, and completing and reviewing the collection of information. Send comments regarding this burden estimate or any other aspect of this collection of information, including suggestions for reducing this burden, to Washington Headquarters Services, Directorate for Information Operations and Reports, 1215 Jefferson Davis Highway, Suite 1204, Arlington, VA 22202-4302, and to the Office of Management and Budget, Paperwork Reduction Project (0704-0188), Washington, DC 20503.

1. AGENCY USE ONLY (Leave blank)	2. REPORT DATE November 2006	3. REPORT TYPE AND DATES COVERED Final Report September 2005 to September 2006
4. TITLE AND SUBTITLE Final Report and Recommendations for Research on Human-Automation Interaction in the Next Generation Air Transportation System		5. FUNDING NUMBERS NA23/DM345
6. AUTHOR(S) Thomas B. Sheridan, Kevin M. Corker and Eric D. Nadler		
7. PERFORMING ORGANIZATION NAME(S) AND ADDRESS(ES) U.S Department of Transportation Research and Innovative Technology Administration John A. Volpe National Transportation Systems Center 55 Broadway, Cambridge, MA 02142-1093		8. PERFORMING ORGANIZATION REPORT NUMBER DOT-VNTSC-NASA-06-05
9. SPONSORING/MONITORING AGENCY NAME(S) AND ADDRESS(ES) National Aeronautics and Space Administration Washington, DC 20546-0001		10. SPONSORING/MONITORING AGENCY REPORT NUMBER
11. SUPPLEMENTARY NOTES This report can be accessed at http://www.volpe.dot.gov/hf/pubs.html.		
12a. DISTRIBUTION/AVAILABILITY STATEMENT		12b. DISTRIBUTION CODE

13. ABSTRACT (Maximum 200 words)

This is the final report of a project to review JPDO documents as they pertain to human-automation interaction, review past system failures in aviation and other contexts involving human-automation interaction, conduct a workshop of JPDO, NASA and academic experts in the area, perform analyses of selected problems, and make recommendations for NASA research needed to support JPDO on these aspects of NGATS. This report first reviews reports issued separately on the failures review and the workshop findings as well as several papers and technical notes. Recommendations for needed research in human-automation interaction are then detailed.

14. SUBJECT TERMS NGATS, human factors, automation			15. NUMBER OF PAGES 37
			16. PRICE CODE
17. SECURITY CLASSIFICATION OF REPORT Unclassified	18. SECURITY CLASSIFICATION OF THIS PAGE Unclassified	19. SECURITY CLASSIFICATION OF ABSTRACT Unclassified	20. LIMITATION OF ABSTRACT Unlimited

NSN 7540-01-280-5500

Standard Form 298 (Rev. 2-89)
Prescribed by ANSI Std. 239-18
298-102

METRIC/ENGLISH CONVERSION FACTORS

ENGLISH TO METRIC | METRIC TO ENGLISH

LENGTH (APPROXIMATE)

English to Metric:
- 1 inch (in) = 2.5 centimeters (cm)
- 1 foot (ft) = 30 centimeters (cm)
- 1 yard (yd) = 0.9 meter (m)
- 1 mile (mi) = 1.6 kilometers (km)

Metric to English:
- 1 millimeter (mm) = 0.04 inch (in)
- 1 centimeter (cm) = 0.4 inch (in)
- 1 meter (m) = 3.3 feet (ft)
- 1 meter (m) = 1.1 yards (yd)
- 1 kilometer (km) = 0.6 mile (mi)

AREA (APPROXIMATE)

English to Metric:
- 1 square inch (sq in, in^2) = 6.5 square centimeters (cm^2)
- 1 square foot (sq ft, ft^2) = 0.09 square meter (m^2)
- 1 square yard (sq yd, yd^2) = 0.8 square meter (m^2)
- 1 square mile (sq mi, mi^2) = 2.6 square kilometers (km^2)
- 1 acre = 0.4 hectare (he) = 4,000 square meters (m^2)

Metric to English:
- 1 square centimeter (cm^2) = 0.16 square inch (sq in, in^2)
- 1 square meter (m^2) = 1.2 square yards (sq yd, yd^2)
- 1 square kilometer (km^2) = 0.4 square mile (sq mi, mi^2)
- 10,000 square meters (m^2) = 1 hectare (ha) = 2.5 acres

MASS - WEIGHT (APPROXIMATE)

English to Metric:
- 1 ounce (oz) = 28 grams (gm)
- 1 pound (lb) = 0.45 kilogram (kg)
- 1 short ton = 2,000 pounds (lb) = 0.9 tonne (t)

Metric to English:
- 1 gram (gm) = 0.036 ounce (oz)
- 1 kilogram (kg) = 2.2 pounds (lb)
- 1 tonne (t) = 1,000 kilograms (kg) = 1.1 short tons

VOLUME (APPROXIMATE)

English to Metric:
- 1 teaspoon (tsp) = 5 milliliters (ml)
- 1 tablespoon (tbsp) = 15 milliliters (ml)
- 1 fluid ounce (fl oz) = 30 milliliters (ml)
- 1 cup (c) = 0.24 liter (l)
- 1 pint (pt) = 0.47 liter (l)
- 1 quart (qt) = 0.96 liter (l)
- 1 gallon (gal) = 3.8 liters (l)
- 1 cubic foot (cu ft, ft^3) = 0.03 cubic meter (m^3)
- 1 cubic yard (cu yd, yd^3) = 0.76 cubic meter (m^3)

Metric to English:
- 1 milliliter (ml) = 0.03 fluid ounce (fl oz)
- 1 liter (l) = 2.1 pints (pt)
- 1 liter (l) = 1.06 quarts (qt)
- 1 liter (l) = 0.26 gallon (gal)
- 1 cubic meter (m^3) = 36 cubic feet (cu ft, ft^3)
- 1 cubic meter (m^3) = 1.3 cubic yards (cu yd, yd^3)

TEMPERATURE (EXACT)

- $[(x-32)(5/9)]\ °F = y\ °C$
- $[(9/5)\ y + 32]\ °C = x\ °F$

QUICK INCH - CENTIMETER LENGTH CONVERSION

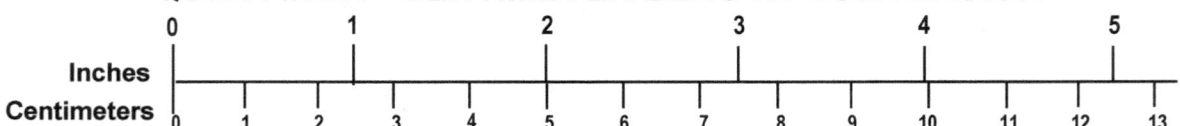

TABLE OF CONTENTS

1. INTRODUCTION 1
   1.1 What is Automation? 1
   1.2 Attributes of NGATS that Affect Human-Automation Interaction 1
   1.3 Abstracts of Reports and Papers Issued Separately 3
2. HUMAN-AUTOMATION RESEARCH NEEDS FOR NGATS 5
   2.1 Highest Priority Research Needs 6
      2.1.1 Task Analysis 6
      2.1.2 Human-Automation Function Allocation 7
      2.1.3 Transition of Authority and Responsibility as a Function of Operational State 9
      2.1.4. Transition from Automation to Human Control 10
   2.2 Research Needed to Support Particular NGATS Operations (by Phase of Flight) 11
      2.2.1 Preflight Planning and Negotiation of 4D Trajectories 11
      2.2.2 Airportal Surface Operations 12
      2.2.3 Tactical Separation Management in En Route and Transition Airspace 13
      2.2.4 En route Modification of 4D Trajectories 15
      2.2.5 Flow and Airspace Reconfiguration Management 15
   2.3 Generic NGATS Design Issues 16
      2.3.1 Human-Automation Performance Metrics and Risk Analysis 16
      2.3.2 Aircrew and ANSP Mental Workload 18
      2.3.3 Aircrew and ANSP Situation Awareness 20
      2.3.4 Aircrew and ANSP Prospective Memory, Decision-Making and Trust 20
      2.3.5 Use of Net-Centric Information 21
      2.3.6 Aircrew and ANSP Displays and Decision Support Tools 22
      2.3.7 Air Crew and ANSP Communication 23
      2.3.8 Simulation, both Human-in-the-Loop and Fast-Time 24
      2.3.9 Information Value and Design Tradeoff Decisions 25
      2.3.10 Aircrew and ANSP Selection and Training 26
      2.3.11 NGATS Management Planning and Policy for the Transition to NGATS ... 27
3.0 REFERENCES 30
4.0 ACKNOWLEDGMENTS 31

1. INTRODUCTION

This is the final report of an 18-month project to: (1) review Next Generation Air Transportation System (NGATS) Joint Planning and Development Office (JPDO) documents as they pertain to human-automation interaction; (2) review past system failures in aviation and other contexts involving human-automation interaction; (3) conduct a workshop of JPDO, NASA and academic experts in the area; (4) perform analyses of selected problems; and (5) make recommendations for NASA research needed to support JPDO on these aspects of NGATS. This report first reviews reports issued separately on the failures review and the workshop findings as well as several papers and technical notes. Recommendations for needed research in human-automation interaction are then detailed.

1.1 What is Automation?

Automation is here broadly defined to include control systems with varying degrees of human supervision, ranging from automatic control that can be overridden to perform management-by-exception, to systems in which the flight crew or air navigation service provider (ANSP) "trades" or "shares" normal control with the automation to varying degrees. We also mean automation to include computer-based information acquisition, trajectory planning, decision support systems, alarm systems and high-level graphic and text displays.

Supervisory control can occur at multiple levels, from no automation of sensing, decision aiding or control implementation to full automation. On a single scale the graded levels might be:
1. The computer offers no assistance; the human must do it all.
2. The computer suggests alternative ways to do the task.
3. The computer selects one way to do the task, and:
4. ---executes that suggestion if the human approves, or
5. ---allows the human a restricted time to veto before automatic execution, or
6. ---executes automatically, then necessarily informs the human, or
7. ---executes automatically and informs the human only if asked.
8. The computer selects, executes, and ignores the human.

Alternatively one may utilize separate scales for differing degrees of automation at the four stages of a functional process: (1) sensing automation; (2) display automation; (3) decision-aiding automation; and (4) control automation. Thus different subsystems or different alternative designs may be characterized at different locations in a three-dimensional space.

1.2 Attributes of NGATS that Affect Human-Automation Interaction

NGATS is expected to embody the following characteristics relative to the current national airspace system (NAS).

- **More Precision** – This refers to tighter tolerances for aviation operations, including the use of time-based metering to achieve assigned 4D trajectories (i.e., position in both space and time), reduced separation, and "equivalent visual" operations. The precision would be enabled by new surveillance technology with increased accuracy, improved weather predictions, improved traffic displays, and other technological advances. Reducing separation and other tolerances, of course, will result in more aircraft to pass through a given airspace in a given time period.

- **More Information and Information Sharing** – This refers to the increased information to be available in the NAS, as well as the sharing of information among all users. More information would enable the more efficient operation of the NAS.

- **More Automation** – This refers to the computer-based technology for planning and execution of air traffic management/air traffic control (ATM/ATC) or in NGATS, air navigation service provider (ANSP) operations. More automation enables taking advantage of the greater precision and more information.

- **Extended Horizon for Planning and Operations** – This refers to planning of ATM/ATC operations over a larger geographic and longer time horizon than today, approaching gate-to-gate optimization. The extended planning horizon would permit more optimization and hence more efficiency and greater capacity.

- **Collaborative Decision-Making (CDM)** – This refers to decisions made through negotiations among users and ANSPs. Collaborative decision-making is to give users a greater role in ATM and to provide more efficiency.

- **Flexibility** – This refers to dynamically changing ATM/ATC airspace configuration and operations in response to changing conditions, such as weather, demand, and security needs. Flexibility enables the more efficient handling of changing conditions in the NAS and hence enabling greater capacity.

These characteristics drive the capacity enhancing benefits to be provided by NGATS. The characteristics also have implications for human-automation interaction in NGATS. Examples of the implications are: with more precision, precise actions need to be taken at precise times (both for safety and to achieve capacity benefits) and there is less time to react to abnormal situations. With more information and information sharing there is more information to access, interpret, and judge whether to use. With more automation, decision support tools will be used extensively. Automation will be involved in planning, executing, and monitoring, and automation will conduct more of the aircraft execution of 4D routes in the air and on the ground. With an extended horizon for planning and operations, humans and automation will need to be aware of conditions (e.g., weather, traffic levels, special use airspace (SUA), etc.) over a larger geographic and time horizon than today. Humans and automation will need to remember to conduct actions farther into the future than today. With collaborative decision-making, there will be more user requests and more negotiations; and with flexibility, controllers and pilots will operate in more, and in changing, modes of operation.

1.3 Abstracts of Reports and Papers Issued Separately

During the project four other reports and two technical notes were issued separately. These are:

1. A Review of Human-Automation Interaction Failures and Lessons Learned (Report No. DOT-VNTSC-NASA-06-01)

This report reviews 37 accidents in aviation, other vehicles, process control and other complex systems where human-automation interaction is involved. Implications about causality with respect to design, procedures, management and training are drawn. A number of caveats and recommendations from the salient literature are discussed with respect to human-automation interaction.

2. Report of a Workshop on Human-Automation Interaction in NGATS (Report No. DOT-VNTSC-NASA-06-02)

This report reviews the findings of a workshop held in Arlington, VA on May 10-11, 2006 to consider needs for research in human-automation interaction to support JPDO. Participants included representatives from JPDO, several NASA centers, and various academic institutions.

3. Human-Transient-into-the-Loop Simulation for NGATS

This is a paper presented at the American Institute of Aeronautics and Astronautics Modeling and Simulation Conference held in Keystone CO on August 21-24, 2006. The paper calls attention to instabilities in tactical separation of aircraft that could occur in NGATS if two conditions coincided, namely (1) control would be continuous in time, and (2) time delays or sample-and-hold control elements occur within any supervisory control loops. The latter are likely because ANSP personnel take finite time to assess situations and make decisions. The paper presents results of a variety of dynamic control simulations using MatLab Simulink® computer-graphic tools.

4. Next Generation Air Transportation Systems: Human-Automation Interaction and Organizational Risks

This is a paper presented at the Resilience Engineering Symposium, Juan-les-Pins, France, November 8-10, 2006. The paper discusses various risks to NGATS safety related to human-automation interaction. Among the issues are:

- Who (human) or what (computer) has authority at what stage of flight?

- What network information should be "pushed", what needs to be "pulled", what should be restricted and to whom?

- Insufficient robustness, reliability and operator trust in decision support tools

- Control instabilities resulting from closed-loop time delays of controller time-sharing of attention, or needed perceptual and decision time

- Inadequate operator mental models of the automation and awareness of what the automation has done, is doing, or can be expected to do

- Lack of a "safety culture" that encourages operator recognition and reporting of errors and near-miss situations

- Inappropriate assumptions of linear transition and scalability as NGATS evolves.

5. A Note on the Possibility of Instability in NGATS Upstream Flow Control to Airports (Technical Note No. DOT-VNTSC-NASA-06-03)

Representation, estimation and optimization of airport capacity are used to establish constraints on the Pareto constraint curve for various combinations of arrival and departure for various time intervals as a function of factors such as weather, runway configuration, etc. The NGATS Concept of Operations 0.2 (JPDO, 2006) poses the following research issue: "With trajectories manipulated 20 minutes or less ahead, how is trajectory stability affected? What is the effect on meeting computed-times-of-arrival (CTAs) and what is the effect on system functions that rely on CTAs?" Measuring aircraft flow upstream to control flow downstream can pose a time delay in the control loop (20 minutes with no prediction, but delay also occurs with prediction as any prediction error amounts to time delay). As pointed out in the paper titled, Human-Transient-into-the-Loop Simulation for NGATS (see above), such delay can cause flow instability if continuous control is implemented. MatLab Simulink® tools are employed to demonstrate the effect.

6. Strategy for Optimum Acquisition of Information (Technical Note No. DOT-VNTSC-NASA-06-04)

This technical note is a brief tutorial on a strategy for optimizing the acquisition of information. It is a procedure well known to decision theorists but hardly understood or applied by those making decisions about spending dollars, time and other forms of capital on the acquisition of information. In the authors' judgment it has wide applicability to NGATS with respect to aircraft and ATM equipage, design of decision support tools, operator training, and system architecture. Acquisition of information refers either to doing research (presumably to discover a state S, the value of some property of an object or event) or designing/deploying some physical instrumentation to measure S. In the latter case, for example, the question may be what instrumentation, if any, to install on an aircraft to provide operating performance and safety of greater worth by knowing S, given the cost of the instrumentation.

2. HUMAN-AUTOMATION RESEARCH NEEDS FOR NGATS

Applied research needs presented here are largely in the form of questions, based on the premise that a well-stated question can have a well-stated answer. The applied researcher's task is to give an answer that is judged credible by and useful to the responsible community (in this case NGATS designers). If a research topic is not narrowed to one or more questions it can elicit many and various responses with few or no criteria for evaluating its applicability to the need or the usefulness of the result.

The research needs are clustered under three first level topical headings: (1) Highest priority research needs; (2) Research to support NGATS operations by phase of flight; and (3) Generic NGATS design issues. A further breakdown is by topical subheadings defining particular research issues.

For each sub-headed topic a short initial statement defines the research issue, and some salient research questions follow. After the list of questions appropriate to that issue, the corresponding "research issue" statements are quoted from the JPDO (2006) Concept of Operations 0.2 . Relevant milestones follow, obtained from recent NASA planning documents (Swenson, Barhydt and Landis, 2006, for the Airspace (AS) program; Young and Quon, ND, for the Integrated Intelligent Flight Deck (IIFD) program). . Note that neither the JPDO issue statements nor the NASA milestones distribute evenly across the proposed research topics, but the chosen topic categorization is the best the authors could manage in order to combine our own proposals with the JPDO issues and the NASA milestones.

Individual questions are not meant to call for separate research projects. Any single research project might deal with several questions, and there would surely be overlap between research projects. Such allocations, especially budgeting, must be left to program managers.

A second caveat is that the proposed research needs are for simulation experiments and analyses quite specifically applied to the NGATS context, and are not basic research. The distinction in Department of Defense terminology is that the recommended studies are in the 6.2 (applied research) rather than 6.1 (basic research) budget categories. This means that the parameters characterizing the human-automation situations (physical systems and environments, procedures, etc.) would bracket those reasonably expected in NGATS. The research should not intentionally undertake an effort to understand human-automation interaction *in general* (e.g., negotiation, function allocation, attention and situation awareness, decision-making and trust, mental workload, situation awareness, human-computer communication, etc.). The research context should map credibly to the NGATS vision(s) so that research results can translate easily into system design recommendations.

It goes without further justification that part of any substantive research task is to determine what research has already been done. Accordingly we do not claim this report to be a review of the literature, except to comment that general reviews of human-automation interaction are available (see General References in Section 3). None of these is adequate to answer the specific and context dependent questions and research needs posed here for NGATS.

2.1 Highest Priority Research Needs

Though the process of development of an NGATS plan is ongoing, and the NASA response to that planning challenge is in its early stages, there are several human factors requirements that transcend the state of the NGATS plan or the stage of the FAA operational evolution plan. These cross-cutting requirements should be considered high priority in making decisions in a research portfolio in large part because of their ubiquity and because they form a basis of understanding human-system performance requirements across NGATS implementation. These are:

- Task Analysis
- Human-Automation Function Allocation
- Methods for Transition of Authority and Responsibility as a Function of Operational Concept
- Transition from Automation to Human Control

2.1.1 Task Analysis

Throughout the process of evolving the next generation airspace plan and undertaking the foundational research to support the operational improvements, a fundamental requirement is to characterize the tasks and functions that are the basis of the evolving system. These tasks and functions are shared and traded among human operators and between human operators and automation in the planned systems. Across all airspace operations or service provisions a required first step is to examine what tasks and functions need to be done and after that to lay out at least a preliminary assignment of who (or what) is going to do them and under what conditions.

Initially, at least, a coarse level task analysis should be performed at an abstract level for all operations -- prior to deciding on allocation of a human or a computer (or an artificial sensor or a mechanical control system) to perform a given function. "Abstract" is this case means that the task analysis specifies independent and dependent variables (measurements, decisions and control actions) qualified by timing and accuracy requirements independent of who or what agent acts.

Tasks should be considered as a hierarchy of goals. A first step is to develop a set of high level goals with the functional processes that support them, and then to determine what lower level goals are embedded within them, possibly iterating to still lower level goals. In this process coordination among functional entities in the work environment should be considered, and their separate goals (and mutual goals) should be considered along with the supportive processes. Processes may affect more than one goal at a time.

A consistent task analysis method should be applied across all NGATS operations. One research requirement then is to adapt the many task analysis task design techniques (Hollnagel, 2003; Vincente, 1999) to NGATS planning. The common task analysis will then form the basis for research in human system operations. Through interaction the human factors researcher will elicit the expert's view of the work domain and how human operators and computers should function within it. This can be done at several levels of detail across the work domain depending on the maturity of the operational concepts.

If and when it is decided that a human should or could perform a task, "cognitive task analysis" should be performed to translate NGATS concepts-of-operation documents into a description of what information the human pilots and air traffic managers must acquire, what decisions they must make, and what actions they must take. This is in contrast to what "operator task analysis" implied decades ago, where doing a job was typically a matter of seeing and doing according to simple and well established procedures, for today the human's task is far more a matter of comprehending patterns of information and making complex decisions.

There is little or no discussion of cognitive task analysis in either the JPDO Concept of Operations 0.2 or in the NASA milestones thus far reported. This is most likely because, while the need for task analysis *is* appreciated at some level of NGATS planning, it has been regarded as a normal planning tool to be done in broad outline but not necessarily in much detail, the detail to be left as an obvious part of later system refinements by design engineers. We assert that task analysis is indeed a research need, to be done in detail by human factors and cognitive engineering professionals skilled in the art. Salient research questions are:

- What existing task analysis methods should be adopted and how should they be combined into analysis techniques appropriate to NGATS?

- What task analyses for NGATS have already been accomplished?

- To what depth should a generic task analysis (i.e., independent of human or computer allocation) be performed?

- To what depth should a cognitive task analysis be performed in the various major tasks? For the aircrew? For ATM personnel?

- To what extent should aircrew, ATM personnel, *and* computer functions be analyzed side-by-side on the same time line or flow chart?

- Should all NGATS task analyses be standardized to the same format?

Obviously, from the above considerations, task analysis cannot wait until system design is in its final stages, but must start in early NGATS planning. Yet it must be iterated as plans evolve and the system becomes more concrete. Therefore the methods for supporting these task analyses will have to be robust at varied levels of definition and detail, for further research depends on the task analysis and the system design, including function allocation between humans and computers and between different people, depends on the research. The initial "abstract" task analysis with hierarchical decomposition will be followed by initial allocation of functions to humans or machines, followed by refined task analysis and definition at lower levels in the hierarchy. As research on NGATS human-machine interaction progresses, task analysis and function allocation must be iterated.

2.1.2 Human-Automation Function Allocation

Although in practice human-machine function allocation (deciding whether human or machine should perform certain sensing, decision or communication functions, and which human or which machine) is closely interwoven with task analysis, it is normally considered a separate topic. Much has changed since the publication of the Fitts (1951) MABA-MABA List (Men-Are-Better-At, Machines-Are- Better-At). Technology has become much more capable of performing sensing, decision, communication and action functions in comparison to humans. Humans are slower, less accurate, and less predictable. Yet under off-normal and unanticipated circumstances machines can look stupid and humans are invaluable in perceiving complex patterns of information, making complex decisions based on probabilistic data and value judgments, and improvising to recover from otherwise disastrous situations.

Rationally, function allocation should result from and not precede task analysis. However there are many functions where the need for human involvement is obvious from the outset. Some important research questions are:

- For what functions is it obvious that the computer will be better, for what functions is it obvious that a human will be better, and for what functions is it unclear, such that task analysis is the necessary determiner?

- For what functions might the human share control with the computer, so that the computer and human work together simultaneously in real time?

- For what functions will the human and machine likely "trade" control, where each acts and then hands off to the other in iterative fashion?

- For what functions should function allocation between human and computer be dynamic? For example, when the human complains that he (she) is overloaded (or when the computer detects human overloading) how might the computer take over? What functions might shift to the human when the computer indicates uncertainty or the human detects the computer to be uncertain or incorrect in it's decision-making?

JPDO Research Issues

What is the appropriate functional allocation between ANSP personnel, flight operators, ground-based automation and airborne automation systems?

What are the responsibilities and boundaries between CM, FCM, TTM, and TSM[1]? Can the same individual perform some or all of them?

How will the flight operators collaborate with the ANSP? How will they be informed of planned routes? How will the information be processed and presented, etc.?

What level of onboard functionality is required for flight crews to safely perform autonomous operations within acceptable workload levels in en route airspace?

How can human flight crews' operational tasks be designed to conform to good human-centered design? (e.g., how much freedom can the aircraft automation exercise without pilot interaction?).

NASA Milestones

In the IIFD program the Tailored Flexible Operator-Automation Management (TFOAM) project element addresses the requirement to perform a task and function analysis beyond traditional simulation studies. The call is for "formal methods analyses [to] allow for a thorough evaluation of allocation concepts beyond traditional simulation studies. The research provides a conceptual framework that can accommodate new technologies as well as new operator roles as they emerge." Milestones AS1.6.01-04 concern safety of

---

[1] CM is Capacity Management, FCM is Flow Contingency Management, TTM is Tactical Trajectory Management, and TSM is Tactical Separation Management.

descent operations and arrival scheduling in the superdensity environment. Milestones AS1.7.01-06 deal with the responsibilities of human and automation for separation assurance, and the potential for pilots to game the system to their own advantage. AS 2.4.04 specifically seeks research on human-automation function allocation. AS2.4.01-11 relates to information management to ensure separation as well as meet user needs. Human-automation allocation is specifically included. IIFD3.1.1 asks for preliminary assessment of functional categories and operational scenarios, and cross-referencing of flight management tasks with functional categories and operational scenarios. IIFD3.1.2 is for analytical human/machine tradeoff studies for functional categories to compare performance scores for both human and machine for each of the cells in the database produced by IIFD3.1.1. IIFD3.1.3 is to define role-and context-dependent function allocation strategies based on factor analyses of performance database and of pilot/operator pair-wise comparisons. IIFD3.1.4 is to develop and implement prototype function allocation schemes based on context detection ability.

2.1.3 Transition of Authority and Responsibility as a Function of Operational State

Methods are needed to understand and execute reversion from an operational state in which automation manages some or most of the airspace operations to conditions in which (because of anomaly, uncertainty, or lack of information) the human operators of the system take over a significant level of responsibility.

An objective in NGATS planning is to design a system that is fundamentally flexible: The Concept of Operations, Version 0.2 states that "a major theme of the NGATS is the emphasis on providing more flexibility and information to users…". That flexibility enables NGATS to consider the user's perspective: " .. the NGATS is also more agile in responding to user needs." With respect to decision and decision authority the purpose of decision-making is: 'required lead times for implementation can be reduced, the responses can be more specific, and solutions can be more flexible to change".

These high level purposes and characteristics imply a system where change is the operating condition, as opposed to a fixed or consistent assignment of roles and responsibilities. The issue of dynamic allocation will surely be raised in particular applications of intended transitions as the design for airspace operations evolves. It is very evident that a high research priority should be assigned across program elements to consider how transition occurs, and how the operating point of the system can be made clear to the local and distributed operators in the system. (Specific instances of this more general issue are encountered in discussion of methods of "push" and "pull" of information). Research should be undertaken across operational and service provision domains to develop a consistent approach to system transition. Prior research suggests the following specific areas of concern:

- Operating values become a dominant issue in the development of flexible systems. Operation must converge on similar valuation processes, and these processes must be kept consistent as the field of operation changes. This requirement holds from conflict alert to schedule determination.

- Mechanisms for switching among tasks at hand and tasks of advisory assessment become critical design elements.

- Team composition changes resulting from operating state changes and impacts of decision aiding are potentially destabilizing.

- Methods of information integration and segregation, and scaling will vary with operating state.

2.1.4. Transition from Automation to Human Control

A special case of flexible operation occurs when the state transition is not based on continued optimization of performance, but rather is based on transition from an optimal automated state to a reduced operating state that is required because of anomaly, uncertainty and/or system failure. Fundamental research could be undertaken across airspace operations to assure that transition from full operating modes to some level of manual control can be designed, developed and tested. This research should be undertaken to assure the feasibility of the planned transitions for safety critical functions. Such techniques are likely to rely on the development and adaptation of mathematical probability analysis (informed by human performance capabilities). Techniques such as those described by Andrews, Erzberger and Welsh (2006) may be adapted.

Automation anomalies or failures can occur at any of the following stages of flight: (1) preflight 4D trajectory planning and negotiation; (2) surface operations; (3) takeoff and climb out; (4) en route tactical separation; (4) flow management; (5) airspace configuration management; (6) descent and landing, especially at superdensity airports. Best methods for recovery are likely to differ from one phase of flight to another. While the need for recovery in the early stages of a system failure may be anticipated and therefore automated operations may continue to some extent, eventually the reduced state may require reversion to almost full manual control.

What the aircrew and/or ATM personnel can or should do in case of anomaly or failure is naturally a strong function of training and procedures. Failure recovery procedures should be evaluated in human-in-the-loop simulations. The hierarchy of alerts, alarms and failure indicators should be analyzed to minimize the burden on the aircrew with respect to telling them about details they can do nothing about.

We believe that a focused human factors research effort in these general areas is critical and should provide a consistent human-automation discipline. Important research questions are:

- How should various anomalies and failures be classified?
- In case of aircraft automation anomalies or failures how should the aircrew be informed?
- In case of ATM automation anomalies or failures how should ATM personnel be informed?
- What aircraft automation anomaly or failure information should be communicated to appropriate ATM, and by what means?
- What means can be employed to permit the aircrew and/or ATM personnel to "buy time" for diagnosis and recovery when anomalies or failures are indicated?

JPDO Research Issues

If the automation fails, what's the backup plan in terms of people/procedures/automation?

What are the responsibilities and liabilities of different stakeholders in the event of automation failure?

2.2 Research Needed to Support Particular NGATS Operations (by Phase of Flight)

2.2.1 Preflight Planning and Negotiation of 4D Trajectories

Preflight negotiations will be accomplished at several stages. From one to many weeks prior to a flight the airline operations centers (AOCs) will negotiate with the ANSP to agree on a 4D en route trajectory. The negotiation will be supported by a shared use of the computerized planning tools, which will somehow participate in the negotiation between airlines, pilots and air traffic management. The 4D en route trajectory will be updated at least once, days to minutes preflight, to (possibly) revise the plan. The overall question is: how will this negotiation take place?

A high priority research issue for negotiation and distributed control systems is how both the state of the planning/negotiation process and the content of current plan are made obvious to the operators in the system. There is also a consistent requirement for explanations to support negotiation and acceptance. Research is needed on how to present both information and process. Specific instances and issues are:

- How should a plan be presented along with the reasons why that plan was chosen, and what procedure should be used to negotiate the plan?
- What limitations does the decision context impose on flexibility for negotiation, e.g., time allowed to come to agreement on the extent of some deviation?
- What happens if the AOC/pilot rejects the computer proposal? How many modification attempts are acceptable?
- Under what circumstances, if any, should a human representing the ANSP step in?
- Can negotiations include monetary or other trades (e.g., of flight trajectory slots)?
- How are automation failures or partial failures identified?

The research should involve simulation with realistic flight plans, subjects who are representative of the pilot, ANSP and AOC communities, and with realistic time constraints. Theoretical knowledge of negotiation and gaming (e.g., Nash solutions, conclusions from "prisoner's dilemma" research) should be brought to bear. Negotiation/decision support tools are important considerations. Some default policies should be proposed as part of the research.

JPDO Research Issues

What level of performance (e.g. precision, integrity, latency, accuracy, etc.) is necessary for the different uses of the 4D trajectories (4DTs), who and where (e.g., operator versus ANSP) should they be performed, and what is the appropriate time horizon?

What is the minimum level of detail for a 4DT? How does this vary according to the type of operation? What gets transmitted between air and ground?

In this more automated process, what is the role the human being plays in the flight planning and negotiation process other than those involved in FCM and TTM?

NASA Milestones

No milestone deals specifically with preflight negotiations, nor is the differing nature of negotiations at different phases of flight (preflight, on the surface, climb out and descent, and en route) detailed in the milestones. Many milestones relate to computer supported optimization or "satisficing", which necessarily includes user negotiation considerations. AS3.4.01-06 includes "determining appropriate roles and procedures that enable users and air traffic service providers" to operate the system, including balancing workload. IIFD1.6.04 is to assess "hazard detection and severity estimation."

2.2.2 Airportal Surface Operations

Although there is pressure to build more runways and make better use of smaller airports located near urban metroplexes, airport surfaces may well be the choke points for NGATS operations. NGATS has key capabilities specifically aimed at the airport surface (i.e. "equivalent visual" instrument approach or "electronic" VFR and "super density" operations) that have implications for automation-human interactions.

What should be the assignments in terms of roles and responsibilities for ramp, ground, and local controllers vis-à-vis the surface computer planner? Should these roles be redefined and reconstituted for NGATS? This is a specific instance of the more general requirement for task analysis and functional allocation. This is made particularly salient in that for superdensity operations NGATS is considering automated airport operations and some remote shared authority among remote controller and local automation. These mixed-mode communications among air traffic controller and flight crews will include voice, data link, imagery and perhaps control commands.

In non-digital modes of interaction how is the system kept informed of human voice induced changes? This is a specific instance of the more general issue of transitions among different modes of operation within the larger system, and what restrictions transition among operational modes (e.g. human to human) will have on system optimization. The research issue is focused on how the different controllers of the system cooperate. Other research questions are:

- Does the aircrew have any options to modify the given plans?
- To what extent will the taxi operation be controlled manually by the pilot, and to what extent automatically through the FMS (presumably with the pilot monitoring and able to take over control)?
- How will information about movement of other aircraft, other vehicles, wake vortices, or errors in movement of own aircraft be communicated to aircrew and/or human controllers?
- Who has responsibility for surface separation?
- How will surface actions be controlled: holds at ramps, on taxiways, runway crossings, or on runways; takeoff clearances; takeoffs?
- Will the human controller or computer intervene to prevent impending collisions?

This research will surely involve analytical modeling and both fast-time and human-in-the-loop simulation methods.

JPDO Research Issues

What new procedures and technologies are needed to deal with current bottlenecks on the airport surface?

How can airport surface and airspace usage be better coordinated?

NASA Milestones

Airportal milestones are not yet available at this writing. Milestones AS3.6.01-06 concern airportal operations, including human cognition and workload.

2.2.3 Tactical Separation Management in En Route and Transition Airspace

Assignment and expectation of responsibilities for separation is critical. Situations where the pilot will self-separate and/or perform station-keeping maneuvers are anticipated. But there will likely also be situations where the ground-based automation will perform separation functions, with control signals, or at least advisory signals data-linked right to the aircraft flight management system (as well as to the pilot when appropriate).

These dual, shared and possibly conflicting roles in separation are of critical priority from a human factors perspective. Prevailing modes of operation (ground centered vs. flight deck centered) and variations of these (station keeping, self-separation in tubes) must each be examined with task analysis. In addition task analyses for transition and failed mode operations should be undertaken. These analyses should then be compared to earlier analyses of free flight and distributed air ground operations to assure that lessons are heeded.

Initial feasibility analyses in tactical separation should be undertaken in system operating modes based on the results of task analyses.. Human-in-the-loop simulations should be used to examine the details of performance.

Since tactical separation management will determine what equipment and automation are required, it should take priority in airspace and airport operations as well as in research in Air Navigation Services and Flight Operations. Specific Issues are:

- To what extent will the climb-out and descent flying of 4D trajectories in poor visibility and in super density operations be controlled manually by the pilot, and to what extent automatically through the FMS (presumably with the pilot monitoring and able to take over control)?

- How much and what information should be presented to the aircrew during these operations?

- How will the computer convey the approach or departure plan to the aircrew (datalink/voice, spatial displays, head-up displays (HUDs), or some combination)?

- How and when will negotiations on the approach plan be conducted?

- How will information about location of other aircraft (particularly for closely spaced parallel approaches), wake vortices, or errors in movement of own aircraft be communicated to aircrew and/or human controllers?

- What are the degrees of freedom for each aircraft in self-separating?

- How is the pilot to be informed that deviation from the accepted trajectory is too great, especially if in conflict with another aircraft?
- How far away from the agreed 4D trajectory may the pilot deviate without being contacted by the ANSP, alerted by on-board equipment, or declaring an emergency?
- Under what circumstances should the controller step in, and by what procedure and communication mechanism does he or she do so?

<u>JPDO Research Issues</u>

What new procedures and technologies are needed to deal with current bottlenecks on the transition airspace?

How is the time dimension of airspace classification to be handled? For example, how far ahead of time will reclassification of a terminal's operation to super density operations be known? How will aircraft that are already in-flight be handled if the airspace is classified to become more restrictive?

What degree of flexibility for self-separating versus managed aircraft is practical when the ANSP is providing TSM?

Is the concept of a Flexibility Volume an effective means to balance operator flexibility against system predictability and flow management?

What is the minimum level of detail for a 4DT? How does this vary according to the type of operation? What gets transmitted between air and ground?

How can performance-based separation criteria based on the overall safety risk rather than using a human-based, fixed standard applied to all aircraft, be implemented?

Can reliable ANSP automation be developed that is capable of providing safe and robust separation assurance without human monitoring or intervention? What level of traffic monitoring is needed on the flight deck when operating in ANSP flow managed airspace where automation is providing separation assurance. What level of vigilance is needed? What capability beyond TCAS is needed?

To what extent does ANSP automation issue clearances without human approval? Can there be, for example, automatic exchange of 4DTs between ANSP automation and the aircraft FMS that are then automatically executed to resolve a conflict?

What are the requirements for a collision avoidance system that is compatible with NGATS tactical separation? Unless mandated otherwise, some aircraft will likely be equipped with legacy TCAS/ACAS systems that may generate unwanted alerts during normal operations; how should this be accounted for?

Can automation become the resolver? Does the human have a role in the separation assurance process? The function of separation assurance is considered best allocated to automation versus ANSP personnel if traffic is highly de-conflicted to avoid human complacency. If research shows this is not the case, ANSP personnel would be required for TSM. What cockpit capability and role for the flight crew is needed if there is no human TSP[2]?

Much research work on arrival procedures invoking airborne spacing or airborne separation has already been accomplished, and detailed application descriptions have been developed through RTCA SC-186 and the RTCA/Eurocontrol. More research is needed to determine how these approach procedures are integrated into super density operations, including ground-based automation support.

Super Density operations will result in many aircraft in close proximity. Consequently an aircraft blundering from its assigned trajectory is much more likely to cause an immediate conflict with another aircraft, and safe avoidance maneuvers may be limited or unavailable. How can super density operations be conducted safety?

Is it practical to allow self-separating aircraft to operate among managed aircraft?

---

[2] The TSP is the tactical separation provider.

NASA Milestones

For "Electronic VFR", what is the minimum airborne separation assurance required for flight in IMC conditions? Is traffic situational awareness good enough, considering that all aircraft must have cooperative surveillance? Will appropriate avionics be affordable enough so that imposing a minimum capability is not an undue burden?

2.2.4 En route Modification of 4D Trajectories

In a user and service based system operation, the issues are: What operational modifications can be made, and by whom? How often are the problems due to transitioning system operating mode? Here again a solid set of task analyses is needed to understand the degrees of freedom in the system and the options for levels of authority and the impact of temporal constraints. Analyses will support some level of definition of what alternatives are feasible. In any case human-in-the-loop simulation studies must capture the variability of the human operators.

Given more information on weather and other conditions and the use of decision support tools, under NGATS there will be more en route negations than today. Negotiations en route are more constrained than those preflight, given the pressure of time, weather and other critical circumstances that may occur. New "planning ahead" and "contingency anticipation" tools and displays should be evaluated in this research. In simulation experiments there will be a strong interaction with tactical separation considerations. Some research issues are:

- How are trajectory changes for weather to be negotiated?
- What will be the roles for the pilot, sector controller, flow control, AOC, tactical computer controller, and trajectory planning computer?
- Who or what has final authority, and how will an end to negations be determined?
- Will communication be via voice, datalink, or a mixture?
- Are there workload implications from the increase in negotiations and modifications and should negotiations be limited during certain periods of workload?
- What are design principles for the different negotiations that need to occur at various times in a flight, e.g., to be quick and easy and for the negotiation results to be safe and efficient?

JPDO Research Issues

To what extent does ANSP automation issue clearances without human approval?

Can there be, for example, automatic exchange of 4DTs between ANSP automation and the aircraft FMS that are then automatically executed to resolve a conflict?

2.2.5 Flow and Airspace Reconfiguration Management

Flow and airspace reconfiguration must work together. A set of reconfiguration policies should be developed so that ATM can be consistent in calling for reconfiguration, and aircrews are not surprised and can know the reasons why.

- How is the aircrew to be informed of airspace reconfiguration?
- How much warning or anticipation is needed?

- Will airspace reconfiguration differentially affect the unequipped aircraft relative to those equipped, and if so what problems might this cause?

JPDO Research Issues

What is the appropriate level of specificity of a 4DT that requires reserving a reasonable amount of airspace without the need for frequent re-negotiation between the aircraft and ANSP, and consequent provision of automation and datalink capability. This trade-off is airspace and density dependent.

What new forms of contingency flow management are required to take advantage of increased tactical flexibility, providing the users with more options while ensuring high throughput and safety? What tools are needed to support the implementation of these traffic management strategies?

With what frequency would airspace configuration changes be made? What is the impact on flight efficiency, ANSP personnel productivity, etc.? What is the appropriate lead time for changing a configuration? How is the dynamic nature of weather incorporated into temporal decisions on airspace configuration?

To what extent are incentives provided or requirements established for "early filing?" How do these requirements affect scheduled operations versus on-demand operations?

How is the time dimension of airspace classification handled? For example, how far ahead of time will reclassification of a terminal's operation to super density operations be known? How will aircraft that are already in-flight be handled if the airspace is classified to become more restrictive?

What air automation is required for corridors? What automation beyond airborne separation is required?

How is time dimension of corridor definition handled, e.g., during creation and dissolution?

At a minimum, how "high-tech" is the real-time update of airspace configuration? Would an ATIS-like voice identification of a configuration suffice, so as to provide information to the widest variety of operators, and thus allow maximum airspace access?

How much control authority is required on the aircraft for onboard CTA to be an effective way of managing arrival flow? That is, how far ahead of time does an aircraft need to receive the CTA to perform to a desired level of control authority, and is that operationally feasible? How far in advance must these airspace definitions be solidified in order to meet charting and automation requirements?

2.3 Generic NGATS Design Issues

This section presents research needs or issues that cut across two or more phases of flight.

2.3.1 Human-Automation Performance Metrics and Risk Analysis

Performance metrics.
Since the viability of the NGATS system relies almost entirely in surpassing unaided human performance limits and reducing the impact of the bottlenecks presumed to be imposed by human performance, a primary set of research issues concerns human-automation performance metrics.

JPDO Research Issues

What new metrics need to be developed in order to evaluate system performance at different levels of detail? How should the results be displayed in order to support decision-making?

What processes and tools need to be developed in order to use such data for effective process control?

What metrics are required to evaluate system performance at different levels of detail? What processes, tools, and data sources are needed for effective and equitable management of system operations?

Do conventional quantitative system performance measures (e.g., information transmission, signal detection, control bandwidth) have a useful role to play?

Safety analysis, including longitudinal spacing and path containment for non-blundering aircraft, needs to specify the blunder monitoring performance required.

Collision risk analysis, including longitudinal spacing and path containment for non-blundering aircraft, needs to specify the blunder monitoring performance required, if any.

Complexity.
The complexity of human-automation interaction in NGATS poses new analysis and risk prediction challenges. Among them is the fact that complexity occurs in many different dimensions and at different scales: Complexity occurs in temporal scales of days to seconds. Physical complexity occurs in distance scales from feet to thousands of miles. Organizational complexity occurs in two-person/automation teams and in national strategic decision centers with dozens of employees, on individual flight decks and in airline operations centers.

- What are relevance and use of metrics of operational complexity, e.g., number of modes of display or control, or number of procedural contingencies, or number and rate of airspace or operating configuration changes?
- How well can these metrics predict system safety and efficiency?

JPDO Research Issues

How do we incorporate uncertainty correctly into a control mechanism for the NAS?

How do we mathematically formulate the system control problem of a system with embedded uncertainty in order to minimize effect of uncertainty, while allowing enough user flexibility, and while keeping the system stable, and minimizing human workload?

NASA Milestones

AS1.4.04 and AS2.4.02 seek to define and cope with traffic and airspace complexity. IIFD2.2.06b is about information uncertainty. IIFD2.2.12 relates to display clutter. AS3.3.03 seeks to understand complexity limits for different classes of airspace as related to operator workload.

Attention Allocation.
Aircrew and ANSP will attend to what they think are the greatest needs, typically a function of the time demands of the component tasks (duration and proximity to deadline) and their judged relative importance. Strategies for aircrew and ANSP attention allocation need to be developed, appropriate to the new or changed roles they will play in NGATS.

- Can the pilot manage multiple alternatives in support of optimization?
- How many options should be presented, and on what basis should the operators' decisions be made?
- How are attention allocation strategies managed in monitoring automation?
- How do the operators know when to attend to optimization, and if they are alerted for input, what form should that alert take?

- What are the default conditions in the optimization process and how does the operator know them?
- How are known biases of over-reliance to be avoided?
- How are fail-soft modes and resilience built into the system? How is reversion to operational levels that can be managed by the human operator assured?

NASA Milestones

Milestone AS1.2.02 seeks to synthesize human factors and operational literature to identify limits of human performance in managing many aircraft. AS2.5.09 considers human performance broadly in the context of separation assurance. IIFD1.8.01 and IIFD1.8.04 are to model operator monitoring requirements and failure modes.

Risk.
Risk analyses should be performed to predict the probability of errors in human-automation interaction. Fault-tree, event-tree and cause-consequence diagrams can be applied to cognitive task analyses, much as has been done in the past to systems such as nuclear power and chemical process plants. A newer approach to human-machine reliability analysis called "resilience engineering" assumes that there will be unpredictable and therefore unavoidable errors, and puts the emphasis on making the system resilient to failure and recovery easy and reliable (see also Sections 2.1.3 and 2.1.4 regarding transitions of control authority).

Spending money on safety is often hard to justify because there is no apparent immediate effect of safety interventions, and if and when accident/incident rates do decline they are often attributed to factors other than the "safety" interventions.

- How best to justify safety expenditures?
- How to decide "how safe" or "how reliable" to engineer a system?

JPDO Research Issues

How does NGATS automation effectively use probabilistic weather information in establishing and assessing risks in decision-making?

NASA Milestones

IIFD2.2.03 seeks to define requirements for the application of predictive hazard models, simulation tools and analysis capabilities to mishap re-creation. IIFD2.2.05 is for automatic identification of the ten most frequently reported safety issues and vulnerabilities revealed through national archives of FOQA/ASAP/ASRS[3] data; results are to be verified by domain experts from operational pilot community and air carrier safety groups. IIFD2.2.7 requires identification and prioritization of events and trends indicating that new flight deck concepts could compromise system-wide safety. IIFD2.2.9 wants to deliver principles for the design of integrated computation, logic and simulation-based prediction tools for mishap re-creation.

2.3.2 Aircrew and ANSP Mental Workload

During normal operations automation in various forms will pose mental workload of comprehending displays and decision aids, keeping track of many control variables,

---

[3] Safety data from the Aviation Safety Action Program, Flight Operation Quality Assurance, and Aviation Safety Reporting System.

remembering and recalling operational requirements, making decisions and executing actions. NGATS will make more information and more precise information available about the operating environment (e.g., weather, SUA, traffic density, etc.); consequently, more negotiations and adjustments may occur in response to changing conditions, and these will affect workload. At very busy times, especially in anomalies involving multiple aircraft, aircrew and ANSP (insofar as they are assigned) must attend to additional inputs, maintain situation awareness under more rapidly changing circumstances, and avoid breakdown due to excessive cognitive workload. In some cases, they may need to "buy time" through clever attention allocation strategies. Some believe that pilots and controllers will not be busier than they are today because of automation, or alternatively, that a high degree of automatic control will never occur. These beliefs must be rejected until or unless research proves otherwise.

- What peak workload situations can be predicted (e.g., from task analyses, from simulation experiments, etc.), and how can they be moderated?

- What workload metrics are appropriate?

- What default conditions should be used when operators become overloaded?

- Under alternative scenarios that vary the degree of automation will aircrew and ANSP be capable of transitioning from low to high cognitive load when unanticipated circumstances demand it?

- What strategies are best for ameliorating mental workload (what is experienced) for a given task-load (complexity of what must be accomplished)?

- What are the workload implications of actions needing to take place at precise times?

- With planning and integration of flight operations across domains (e.g., surface transition airspace, en route), will more cooperation be required among controllers responsible for the different domains and sectors, and will this affect workload?

JPDO Research Issues

How will increased automation and new technologies impact flight crew and ANSP workload? Items to investigate in this research include: Are there cockpit workload limitations that inhibit the proposed technologies?

What impact do the changing roles and responsibilities (flight deck vs. ANSP, automation vs. human, etc.) have on safety?

What impact does the changing workload and changing workforce have on safety?

NASA Milestones

Milestone AS1.2.02 seeks to synthesize the human factors and operational literature to identify limits of human performance in managing many aircraft. AS2.5.09 considers human performance broadly in the context of separation assurance. IIFD1.8.01 and IIFD1.8.04 are to model operator monitoring requirements and failure modes.

2.3.3 Aircrew and ANSP Situation Awareness

Aircrew and ANSP (insofar as they are assigned) must monitor and maintain situation awareness over long and boring periods of nominal operations under automatic control (with a possible need to impose activities for the purpose of maintaining alertness). Given the dynamic, responsive, and information-intensive nature of the operation of the NGATS systems, there is a need to support and assure the necessary levels of awareness of the operating state of the system, awareness of individual areas of responsibility, and awareness of the impact of individual actions on the system as a whole. Research would seek to extend the situation awareness paradigms from individual to collective awareness.

Unfortunately situation awareness is not well understood, and there is sparse theoretical underpinning in a form that is quantitative, predictive and therefore usable by design engineers.

- What situation awareness metrics should be used?

- Can automation effectively observe/measure situation awareness (or inattention or distraction) and call attention to salient and critical events contingent on such measures?

- Will the aircrew and ANSP be capable of knowing the operating state of the system and the rules, operational concepts, and actions that automation and/or other people are currently using to control it?

- How might automation remind the aircrew and ANSP of the current operating state and what response is appropriate when and anomalous situation arises?

NASA Milestones

Milestone AS2.2.02 intends work with industry and the JPDO to model situation awareness in control, navigation and surveillance. AS2.5.01-03 are aimed at getting input from service providers on operational concepts and to improve situation awareness and human-automation functional analysis milestones AS3.5.06 seeks to establish human-automation cooperation in separation assurance. IIFD1.1.11 and IIFD1.1.15, which deal with measuring operator state, and by inference an operator's ability to know the system state, and especially to recognize hazardous states. IIFD1.8.04 deals with operator monitoring requirements. IIFD2.2.06b relates to displaying information uncertainty. IIFD2.4.03- 09 are concerned with sensing and monitoring operator state. IIFD2.2.10 specifically seeks to model situation awareness. IIFD3.3.02-07 are for sensing and display of hazard information to improve operator situation awareness.

2.3.4 Aircrew and ANSP Prospective Memory, Decision-Making and Trust

Prospective memory is the ability to remember task demands and information required for future task actions in spite of intervening task demands (i.e., where stimuli and appropriate responses for a task are interleaved with stimuli and appropriate responses for other tasks). This is a serious problem in multi-tasking and supervisory control and it is exacerbated by task-load. NGATS will involve longer time horizons for planning operations and will involve planning across domains (e.g., surface, transition, and en route). These imply remembering planned task actions further into the future.

- In which NGATS situations is prospective memory likely to be a problem?

- What means are available to enhance prospective memory?

- In routine tasks, decision-making is perfunctory and reflexive (skill- and rule-based), but in off-normal situations routine decision criteria can lead to trouble and, therefore, a higher level of cognition (knowledge-based decision-making) is called for.

- What are the cognitive implications of managing automation "by exception"?

- Trust is not an idea with a long history in human factors engineering, but is regarded as a critical factor influencing the use and misuse of automation and decision support tools. Can a "best" level of trust be established for each component of automation and for all the information elements that will be available?

- How can an operator's level of trust be measured?

NASA Milestones

Prospective memory is related to attention, situation awareness, workload, complexity and other factors named in the milestones as covered, for example, by AS2.6.07, particularly with regard to superdensity operations. IIFD1.5.11 is explicitly concerned with task demands. Milestone IIFD3.2.2 is to validate flight deck guidelines and information and display requirements through assessment of usability, acceptability, suitability, and safety of first generation adaptive display and interface technologies.

2.3.5 Use of Net-Centric Information

A distinguishing feature of the planned NGATS is so-called net-centric information, which is intended to be broadly available via digital communication links to humans in the system. Some of it will be real-time, and some stored or buffered at various ground or airborne locations. This includes ADS-B surveillance communication as well as other datalink communications between aircraft and ground. The US Department of Defense has pioneered net-centric information systems, but application to NGATS poses many new research questions, listed below. Many can be asked specifically for a particular operating state of the system (e.g., the state resulting from performance-based services, self-separation or ground-based separation, classic arrival or super density operations, etc.), as well as in general. We place the research questions in this section because they are cross-cutting and should be investigated with respect to each phase of flight.

- What information should be "pushed" and what information should be "pulled"? What information should necessarily be presented to the aircrew and/or the ANSPs, and what information should be available only by being accessed at their will?

- How should aircrew and air traffic managers know what information is available on the information network?

- How will they know whether to expect it to be pushed or pulled?

- How should they be alerted as to when they should access information, or when frequently used information has been updated?

- How will information available to be pulled be organized and accessed?

- Should NGATS use keywords as in Google®, some hierarchical system like a conventional library, or functional contingencies, e.g., based on the airspace in which the access is being undertaken?

- How should pushed information be organized?

- What should be the tradeoff between time-criticality and importance? Can "information value" decision analysis (Sheridan, 2006) be fruitfully applied to this problem?

- What information should be unavailable to what person or computer?

- Should operators in the system be able to regulate levels of detail or quantity of information?

- On the integrated flight deck, should datalinked information be displayed in a common place, or distributed among existing aircraft displays?

- How much information can be absorbed for particular decisions and negotiations?

JPDO Research Issues

How much network flexibility is practical from an economic perspective (e.g., connectivity) and from an operational perspective (e.g., local knowledge training requirements, etc.)?

How do we manage information overload by use of 'intelligent decision support systems' and context sensitive information? What information should be pushed vs. pulled?

NASA Milestones

Various milestones concerned with information requirements relate to all of the net-centric information issues identified by the workshop. IIFD1.4.02 concerns display formats and media to achieve visual/ flight deck information requirements. IIFD3.2.2 deals with information needs, while IIFD3.2.3 is about how to display it. The question of what is "pushed", what is "pulled" and what is restricted from some persons in the system is not emphasized in the NASA plans.

2.3.6 Aircrew and ANSP Displays and Decision Support Tools

Displays range from alarms to warnings and from continuous flight control displays to advisories and decision support tools. Major advances in computer-based flight deck displays occurred with the glass cockpit and the electronic flight bag, but must be extended in NGATS because of additional pilot responsibilities. Some decision support tools have already been implemented in today's NAS, but the changed role for (some) ANSP personnel will require display systems that go well beyond what are now in use.

- What decision support tools are appropriate for pilot or ANSP decisions in (1) preflight trajectory negotiation, (2) taxi, (3) takeoff, (4) climb-out, (4) separation, (5) flow control, (6) aircraft maintenance, (7) en route trajectory renegotiation, (8) airspace reconfiguration, or (9) descent and landing operations?

- What are design principles for these decision support tools, which will be much more common, so that they are safe and easy to use with confidence?

- How can the design of the many decision support capabilities, some of which may be used infrequently, aid operator memory for how to use them? For example, should there be some standardization of the decision support tools?

- How can the output of decision support tools be designed so that it is easily comprehended and prompts the users to remember how to execute their actions?

- What display standards should be developed (in cooperation with avionics and airframe industry partners?

- How should the pilot or controller be helped to anticipate dynamic changes in airspace configuration and understand the reasons for it?

- How should ANSP instructions or net centric information or other information be "tagged"?

    o Should it indicate relevance to normal operation or emergency procedure, especially with respect to recent changes in the system?

    o Should it indicate current estimate of accuracy or reliability?

    o Should it indicate trends as well as predicted values, including predicted accuracy or reliability?

    o Should it specify who is expected to act on the information and/or who has authority to act?

JPDO Research Issues

What capabilities in aircraft flight management and decision-support are needed for trajectory-based operations?

What are the appropriate contents of the flight object?

How do we manage information overload by use of 'intelligent decision support systems' and context sensitive information?

What info should be pushed vs. pulled?

NASA Milestones

IIFD1.4.02 and IIFD1.4.06 are about display formats to meet information needs, essential to in-flight negotiations and decision-making.

2.3.7 Air Crew and ANSP Communication

A major change in the NGATS paradigm relative to the current NAS is the free flow of many types of information from many sources to support human-automation control. Therefore the issues of human-human and human-automation communication are very critical. While a substantial literature already exists on datalink, NGATS will depend much more on communication.

- Under what circumstances should voice be used in NGATS? Under what criteria should datalink be used?

- How are human-to-human and human-to-decision-support-tool communication/coordination conducted?

- If they are conducted in a mixed voice and digital mode (through either textual or visual images), how are the procedures and protocols of that interaction to be managed?

- What special terms or procedures will assure consistent and accurate coordination?

- How important is the current "party line" (same voice frequency) of shared information for situation awareness, and need this be continued in NGATS, or can its advantages be replaced in some form on datalink?

- How will typing speeds and typing errors of pilots or air traffic managers affect datalink communications?

- How will visual attention demands of datalink communication on pilots or air traffic managers suffer from other visual attention demands placed on them?

NASA Milestones

IIFD2.2.01 and IIFD3.2.02 seek to specify flight deck information requirements. Examples of milestones aimed at specifying information requirements are mentioned in many of the previous sections.

2.3.8 Simulation, both Human-in-the-Loop and Fast-Time

This research issue is about the means for conducting human factors research, specifically simulation. Simulation is the major technique both for identifying operational problems and for refining and verifying procedures, systems and training. In most simulations actual human operators will interact with each other and/or with the appropriate procedures and technology to perform assigned tasks.

Human-in-the-Loop Simulations
Most of these simulations will be conducted with credibly trained and experienced pilots or controllers; the simulated environment and automation technology will be embodied in computers.

JPDO Research Issues

How is the safety and effectiveness of adaptive automation ensured, to what extent must automation actions be pre-tested before put into operational use?

How does this differ between strategic/informational roles versus tactical/prescriptive roles?

NASA Milestones

AS.1.5.04 is to develop methods for quantifying the safety level of human operators in the presence of uncertainties in ANSP systems, using simulation for validation. AS.3.5.06 expands human-in-the-loop simulation of service-provider-based automated situation awareness to focus on controller/pilot roles and responsibilities for time-based metering with automated separation assurance, including failure and recovery models, mixed equipage operations, and human/machine allocations. AS.4.5.01 is for simulation analysis of service-provider-based automated separation assurance with complex traffic, metering, hazardous weather, and failure recovery.

Caveat on realism.
Because high-fidelity human-in-the-loop simulations can be very expensive to set up and run, it is recommended that initial simulations be just realistic enough to provide the subjects a good idea of the task. Much can be learned from such crude simulations. Eventually, of course, high-fidelity simulations and piloted testing with actual systems must be performed to verify performance and safety.

NASA Milestones

Milestone IIFD3.2.2 assesses initial multi-modal presentation formats and interaction methods for 4D plus uncertainty display concepts and virtual visual environments. IIFD2.4.7 calls for a low fidelity simulation study to investigate operator engagement indices for different levels of human/automation integration.

Fast-Time Simulations.
Fast-time simulation experiments can be conducted where computational models of human operators are sufficiently well developed and robust, and where operator models can be connected to computer models of the automation, aircraft, etc.

NASA Milestones

Milestone AS.2.7.01 is to develop a method for modeling human workload in fast-time simulations and validate models against actual workload measurements. IIFD1.5.7 will use validated models of attention allocation and prospective memory to develop error mitigation strategies. IIFD1.8.1 will develop models of distributed operator/automation systems, including definition of desired/required safety properties. IIFD1.8.4 should provide formal models of operator monitoring requirements, including abstractions of operator "failure" modes. IIFD1.8.6 is to define/refine computational models for prediction of human-automation integration vulnerabilities of flight deck technologies. IIDF2.2.2 is to develop models to analyze the effects of operational conditions, task demands, and organizational policies and procedures on vulnerability to pilot errors and accidents in the NGATS highly automated environment.

Role-Playing Simulations.
Before programming computers to implement even crude simulations it is sometimes useful to utilize role-playing games. For example in the "Wizard of Oz" technique humans play the role of the computer, responding to control decisions or information queries, by indicating system responses or providing requested information. In NGATS development such role playing is probably best used at early stages.

2.3.9 Information Value and Design Tradeoff Decisions

Information value research is concerned with the tradeoff between (a) increased safety and efficiency from providing more information and (b) the cost of providing that information. In other words, by knowing how to respond under each and every situation that occurs, better decisions can be made. However that "detailed and usually anticipatory knowing" (over and above having to commit to action knowing only the likelihood of the various situations) comes at the cost of providing that information

- Can the value of control actions taken with certain knowledge of measured states be estimated, and compared to the value of control actions taken in partial ignorance (with only probabilistic knowledge)? How well can the probabilities of anomalous states be estimated?

- If continuous computer-based decision-making substitutes for human controllers who in the past have been positioned in "outer" control loops to control aircraft trajectory, there is the possibility of time delay or intermittent sampling because of computer multi-tasking. (The "inner" loop is the pilot/autopilot controlling the aircraft in pitch, roll, yaw and airspeed). This poses a danger of dynamic instability. Delayed outer loop decision-making might occur, for example, when sectors become overloaded, or when airspace is reconfigured (Sheridan, Corker and Burki-Cohen, 2006).

- How can "outer-loop" instability (e.g., in maintaining separation, or in flow control) due to human observation and decision delays be prevented?

JPDO Research Issues

How to transform the aircraft equipage paradigm (cost, time, level of integration in the flight deck, etc.) to reduce both the cost of adding advanced capabilities in the aircraft and to reduce the time it takes to upgrade aircraft as capabilities evolve?

What is the cost benefit and safety assessment of ground-based versus airborne conflict detection/resolution automation? What information displays and alerting does a pilot need to safely oversee conflict detection and resolution when there is no human on the ground responsible for tactical separation?

With trajectories being manipulated at a 20 minute or less look ahead, how is trajectory stability impacted? What is the impact on aircraft keeping CTAs, and what is the impact on system functions that rely on the CTAs?

What level of flexibility in airspace structure (e.g., routes and boundaries) is needed to achieve operational goals, including efficiency, capacity, and environmental goals? To what extent can the selection of predetermined structures support operational needs? How should flexible routes be defined and displayed to flight crews and ANSP personnel?

Is developing the capability for aircraft to perform autonomous operations economically justifiable? Is there a sufficient business case for enough users to equip? There needs to be a sufficient total benefit for this capability to offset the cost of developing and equipping with the capability.

Is it cost effective to build the capabilities to do both ANSP Flow and Autonomous Operations?

NASA Milestones

Instability is a category of hazard. Milestone AS3.3.03 seeks to evaluate the level of traffic that can be safely handled in a sector. AS3.3.04 is concerned with the frequency of airspace reconfiguration and its effect on operators.

2.3.10 Aircrew and ANSP Selection and Training[4]

A current wave of retirements of air traffic controllers and managers is posing serious challenges regarding the selection and training of future FAA air traffic personnel. Probably some modification of selection criteria and training should soon begin in view of the increased amount of automation and the changed role of air traffic management personnel in the NGATS 2020-2025 time frame, even though that remains 14 years away. Salient research questions are:

---

[4] Responsibility for training for NGATS may not reside with NASA. It is included here for sake of completion and in consideration that JPDO and FAA may wish to refer to this document.

- What should be the technical background criteria for NGATS ANSPs?

- Will the normal computer and automation experience of the new prospective controller generation adequately prepare them, or must selection criteria be made more stringent with regard to formal technical education?

- What new or different training methods should be developed for NGATS controllers and flight crews, given that the controller role will change toward monitoring, management by exception, and supervisory control, and away from vectoring individual aircraft? On-the-job training may be inadequate.

- How to train, and at what appropriate level, for understanding of the automation systems including failure detection?

- How to design simulator-based refresher training with random failures and anomalies?

- Should there be different training for controllers manning different domains of airspace?

- Should ANSPs rotate through serving these different domains so that skills in vectoring individual aircraft are maintained (in case NGATS automation goes down)? Should the controllers' training be organized into functional categories such as high altitude, high automation, or transition airspace?

- Should crew resource management be extended from the flight deck to the controller community?

- Is enlarging the "team" to include both pilot and ANSP impractical due to different organizational loyalties?

- How to train an understanding of what the automation is supposed to do or can be expected to do, what it is doing now, or probably will do next.

NASA Milestones

There appears to be little emphasis on "understanding the automation" in the current NASA research milestones, though that can be considered part of situation awareness as covered by several of the milestones cited above. Perhaps that issue is relegated to FAA pilot and controller training. Milestone AS.2.7.02 is to develop predictive, conceptual-level, safety assessment method for ill-defined complex interacting systems (including the NAS). This includes procedural training. IIFD1.3.01 dea's with understanding operational hazards. IIFD1.5.04 concerns crew training protocols and validation of operational protocols and crew training guidelines for the integrated flight deck.

2.3.11 NGATS Management, Planning and Policy for the Transition to NGATS[5]

This transition poses many challenges. NGATS cannot possibly come into being all at once. Development, proof-testing, and FAA approval of various systems will occur in stages. For economic and system availability reasons, different air carriers and general aviation operators will install different NGATS compatible equipment at different stages and times. Somehow the system will have to adapt

---

[5] Responsibility for management, planning and policy for transition to NGATS may not reside with NASA. It is included here for sake of completeness considering that JPDO and FAA may wish to refer to this document.

to this evolution over the next two decades. Among the human-automation interaction issues that occur due to transition are:

- How to modularize NGATS subsystem development and certification in bite-size increments?

- How much human-in-the-loop fidelity is necessary at each level of development?

- Should a suite of human-automation interaction options be maintained until much or most of the NGATS development is completed?

- What organizational/cultural impediments must be overcome in the transition to NGATS?

- How to evolve to a safety culture that encourages reporting of errors and inefficiencies, with concurrent vertical collaboration, in contrast to the current adversarial arrangement?

- What effect will co-location of en route centers and TRACONs have on system operations?

- How to maintain participation by the human factors community to provide the system architects and designers with timely design requirements in an understandable language?

- How are equity issues between aircraft (e.g., ones that affect time, fuel, and turbulence considerations) to be resolved at various stages of the transition?

- What humans should be participants in the decisions (pilot, sector controller, AOC, flow control, tactical computer, 4D trajectory planner)?

- Under various credible scenarios, who has responsibility for overall system safety at different stages of the transition?

- How to design for scalability, i.e. the robustness of the system to handle fewer or greater numbers of aircraft, smaller or larger aircraft, fewer or greater numbers of airports and runways, etc? What are appropriate benchmarks and constraints on scalability from a human factors perspective, i.e., how will scaling affect attention and workload, situation awareness, and memory?

"Systems of systems", while a buzzword that is debunked by many, has the serious implication of combining models from different disciplinary sectors, for example evolving quantitative or even semi-quantitative models that usefully combine aerodynamics, human-automation decision and control, and economics. NGATS development and efficient, cost-effective operation will see many such demands, and the current art of such modeling is in its infancy.

JPDO Research Issues

Can automation ever be 'responsible' for separation assurance or is the human (either Flight Operator or ANSP personnel) ultimately responsible?

How will legal liability issues be dealt with for self-separation?

To what extent can national policy on access to airspace be transparent and implemented within automation?

What incentives and consequences are applicable to encourage stakeholders to exhibit desired behaviors?

<u>NASA Milestones</u>

As suggested earlier, research on management of project design to accommodate transition is important, but is not emphasized in the NASA planning documents. Scalability is a problem of managing system design and redesign over time. It does not appear to be emphasized in current NASA milestone definition.

3.0 REFERENCES

3.1 General References

Degani, A. (2004). Taming HAL: *Designing Interfaces Beyond 2001.* New York: Palgrave MacMillan.

Dekker, S., & Hollnagel, E. (1999). *Coping With Computers in the Cockpit.* Altershot, England: Ashgate.

Sarter, N. B. & Amalberti, R. (Eds.). (2000). *Cognitive Engineering in the Aviation Domain.* Mahwah, NJ: Erlbaum.

Sheridan, T.B. (2002). *Humans and Automation,* Santa Monica: Human Factors and Ergonomics Society. New York: John Wiley.

Sheridan, T. and Parasuraman, R. (2006). Human-Automation Interaction, Chapter 2 in R. Nickerson (Ed.). *Reviews of Human Factors and Ergonomics, Vol. I.* Santa Monica: Human Factors and Ergonomics Society.

3.2 Cited References

Andrews, J., Erzberger, H. and Welsh, J. (2006). Safety Analysis for Advanced Separation Concepts. *Air Traffic Control Quarterly, 14*, 5-14.

Fitts, P.M., (Ed.) (1951). *Human Engineering for an Effective Air Navigation and Traffic Control System.* Washington, DC: National Research Council.

Hollnagel, E. (Ed.) (2003). *The Handbook of Cognitive Task Design.* Mahwah, NJ: Lawrence Erlbaum Associates.

Joint Planning and Development Office (2006a). Concept of Operations, Version 0.2, <Techhangar.JPDO.Aero>

Sheridan, T.B.(2006). *Strategy for Optimum Acquisition of Information.* Technical Note TN-DOT-VNTSC-NASA-06-04. Cambridge, MA: Volpe National Transportation Systems Center.

Sheridan, T.B., Burki-Cohen, J and Corker, K.M. (2006). Human Transient into-the-loop for NGATS. American Institute of Aeronautics and Astronautics, Modeling and Simulation Conference, Keystone CO, August 21-24.

Swenson, H., Barhydt, R. and Landis, M. (2006). Next Generation Air Transportation System (NGATS) Air Traffic Management (ATM)-Airspace Project. Version 6.0. National Aeronautics and Space Administration.

Vicente, K. J. (1999). *Cognitive Work Analysis: Towards Safe, Productive, and Healthy Computer-based Work.* Mahwah, NJ: Erlbaum.

Young, S. D. and Quon, L. (ND). Aviation Safety Program: Integrated Intelligent Flight Deck. Authors.

4.0 ACKNOWLEDGMENTS

Dr. Richard John of the Volpe Center was instrumental in initiating this project and contributing ideas throughout. Dr. James L. Poage of JLP Performance made a number of valuable suggestions. Dr. John Cavolowsky, Deputy Program Manager for NASA Airspace Systems, served as Technical Monitor.

www.ingramcontent.com/pod-product-compliance
Lightning Source LLC
Chambersburg PA
CBHW081806170526
45167CB00008B/3352